CYBERSECURITY
FOR EXECUTIVES, MANAGERS, AND LEADERS
ARCANE KNOWLEDGE OF THE
DARK ART
FOR SUPERIOR DECISION MAKING
THE CYBER KILL CHAIN
DENNIS HOUCHIN

CYBERSECURITY

FOR EXECUTIVES, MANAGERS, AND LEADERS

The Cyber Kill Chain

Dennis Houchin

Ad Hoc Press
P.O. Box 20
Gladeville, Tn 37071

Publisher's Cataloging-in-Publication Data
Houchin, Dennis, 1961 –
p. cm
Includes bibliographic references and index.
ISBN: 978-0-9861205-7-2

1. Cybersecurity. 2. Information System Management.
I. Houchin, Dennis, 1961-. II. Title

ISBN: 978-0-9861205-7-2
0707
http://DennisHouchin.com

Dedicated to
all the lovely people who do
all the lovely things in
all the lovely ways

And who improve the world
with every breath they breathe
through all their living days.

Table of Contents

PROLOGUE

It's late winter in 1980 and a light snow is beginning to fall. I'm in Arizona driving down the switchbacks on my way to Sedona. Fleetwood Mac is in the cassette player and I might be going faster than I should have, but not as fast as I could have.

Having recently gone through a course of study at the College of Business at Northern Arizona University in data processing, I am entering into the professional workforce. It might be faster than I should have, but not as fast as I could have.

I am a computer programmer entering the business world for the first time. My previous jobs have been mostly blue-collar or no collar. Now I'll be studying with an experienced senior programmer and learning how to create custom software for businesses.

The programmer I am to work with is my first mentor. His name is Felix. Even in 1980 Felix is considered an old school programmer. Felix disdains anybody that, well, isn't him. He knows how to program computers and how to follow the instructions necessary to satisfy customer requirements. And he's the only game in town when it comes to this particularly unusual programming language. Which makes him extremely valuable. For a while.

After a short tutelage with Felix, the owner of the company saw enough of my programming skills and ability to relate to clients, that he now invited me to go with him to open a new office in Tucson. And just like that I am now the Senior Programmer for a brand new office in a brand new market. Only later would I learn that Felix would forever resent me for taking a job he felt he deserved more.

The early 80's are an exciting time for the computer industry. Many businesses still rely on manual bookkeeping processes. Many businesses are hesitant to trust their important records to a computer system that seemed magical to them.

I don't yet realize that I'll be spending much of the 80's working directly with managers in small businesses, from front line supervisors to department managers, to CEO's to Business Owners. Instead of spending day and night programming computers, I'll have to listen to people I'm not comfortable being around. And not just listen to them, but help them understand how technology could benefit their company.

Although I am becoming experienced at this somewhat unnatural colloquy, it is never easy, and I am becoming frustrated that my results were not as consistent as I'd like.

Fortunately, my patience is well developed by this time. And an answer would eventually present itself. An answer that brings about some incredible changes as I follow a path that leads me into Cybersecurity.

1

THE BIRTH OF AN IDEA

Today it's early January in 2009 I'm sitting on a Lanai in a Condo on the Kona Coast of the Big Island of Hawaii. The bone chilling Flagstaff winter of 1980 drifts quickly across my mind, then melts away as the sun rises over the waves crashing on the rocks twenty feet below me. The rest of the family is sleeping while I work on an application for a trademark.

It's the fifteenth anniversary of opening my business as an independent technology consultant. But that milestone is far from my mind. Today I am focused on the future and a breakthrough idea that has recently gelled in my mind.

After several years of gestation the idea has finally been born, fully matured!

In 1994, when I created my business, Ad Hoc Information Systems, I already understood that technology was really the least of my challenges. In fact, the business name was derived, in part, from the acronym for Analysis and Design for Human Oriented Communications.

I knew even then that the challenge was using technology to provide meaningful information to people that didn't necessarily enjoy technology, in the way that geeks like myself did.

I had spent several years informally studying personality types, particularly those based on the work of imminent psychologist and pioneer in the field of personality type, Carl Gustav Jung. Many of the current popular authors on personality types use the vocabulary and vernacular and theories that were invented by Carl Jung. My interest in personality type was focused on its application to work relationships, particularly as they applied to technogeeks and communicating more effectively with management.

Over and over I witnessed the same phenomenon, a manager, who was typically logical, professional, comfortable around others, and good at listening, would visibly change in the presence of a technogeek. Their patience would grow short. Their eyes would glaze over and their discomfort would become palpable.

I imagined that instead of hearing the words that were coming out of the technogeeks mouths, they were simply seeing potential dollar signs float out of their mouths. Dollar signs that indicated a reduction in profit.

Of course it's the job of management to control the utilization of resources to meet important metrics, many of which are preceded by dollar signs. Whether the management task involves people, priorities, processes or products, there are key metrics that will relate directly to profit. So it's natural that cost is among the principle lenses through which management will view any discussion of technology.

On the other hand, technogeeks view their projects through lenses that are very different from management's lenses, chiefly focused on quality or 'elegance' of the solutions that will address the technical and logistical concerns of the company. Their natural tendency is to be more dismissive of cost concerns.

Technogeeks also tend to communicate using a vernacular that, while common to them, is foreign to those who have spent their careers excelling in management and financial or other governance issues. This specialized vernacular, while facilitating communication with those that are familiar with specialized technical vocabulary, are actually barriers to effective communication.

I've seen the following scenario happen many times. A technogeek will be explaining a proposed security solution to a complex problem to a group of managers. After only a few minutes of listening to the nervous technogeek, who is uncomfortable talking to people, the managers will begin to show signs of going into a trance. Their eyelids grow heavier and heavier, their breathing becomes slower and deeper, in.... and.... out...., their arms and legs grow heavier and heavier and growing more and more difficult to move.

When the technogeek finishes, alertness returns to the managers who have predictable questions: How much will it cost and how long will take to implement? How secure will this make us - can you guarantee this will keep us safe? What's the worst that could happen if we don't do this?

While these are important questions to answer, I believe it is equally import that managers have a better understanding of a few of the details of technical solutions so that they are able to make higher quality decisions when it comes to IT and particularly IT Security issues.

It is for the benefit of managers and executives whose decisions will ultimately impact the security posture of the organization that this book is written.

This book will present an in depth overview of cybersecurity that will provide sufficient background information so that decision makers will no longer have to fight the loss of consciousness during an IT Security Presentation and the decisions they make will be better informed and more likely to have a positive effect on the bottom line.

At the same time, I will present sufficient information for technogeeks that will enable them to communicate their messages more effectively to management.

By doing this, I believe this small work can be an effective tool that will help improve the security posture of companies throughout the world.

Most of the content that follows has been extracted from a live training that I provide to organizations. The live/virtual training provides their management with the opportunity to experience many of the tools discussed throughout this work on their own systems. This is typically more effective than watching someone else use the tools.

The ideas explored herein have been developed over a 40 plus year career in Information Technology.

This volume, the first of the series, will address the Cyber Kill Chain in an entertaining format. One that may also be educational.

2

THE STATE OF THE ART

TL;DR - A SCARY CHAOTIC MESS

The field of Cybersecurity is one of those rare fields where the discovery of more answers and the development of greater tactics, techniques and procedures leads directly to a greater distance between our current state of security and an idealized state of perfect security.

There's no question that people that choose a career in the field of cybersecurity are both intelligent and diligent. The intelligence and diligence on the part of cybersecurity professionals is, however, matched by the intelligence and diligence of the opposing antagonists. The biggest difference between the two groups is that the opposing antagonists are rarely constrained by an ethical code that is consistent with professional ethics in most areas of legitimate commerce.

This is not to say that many of the antagonists are not constrained in some way by a moral or ethical code. Hacktivists often do their work in accordance with some sort of ethical code, it just isn't what most of us would recognize.

The first computer hackers were actually explorers. They looked at hardware or software systems and asked, "What else can this thing do?" They would then set about the tedious task of finding all of the possible answers to that question.

Those early hackers were responsible for some of the earliest innovations in technology, from portable computers to car phones to using a mouse for input.

But like any new discipline, it didn't take long before there were a mix of personalities and a mix of motivations involved. Some were interested in improving systems, others were more focused on finding vulnerabilities that could be exploited.

Hackers whose tactics, techniques, and procedures predictably and inevitably lead to financial, corporate, or personal harm are known as Black Hat hackers, while those who use the same tactics, techniques, and procedures to protect people and organizations are known as White hat hackers. Those who engage in both legal and illegal hacking are known as Grey Hat hackers.

As the computer industry blossomed and companies such as Banking and other Financial Institutions began using technology to process their transactions, they simultaneously became rich targets for hackers.

As individuals and merchants began using computers for their financial transactions and as money in the form of credit and debit cards began flowing through Point of Sale systems, the target environment for attackers with a financial motive grew exponentially.

More and more hackers discovered there was indeed plenty of money to be made by utilizing their technical skills, to attack specific financial systems directly in order to siphon off funds through barely noticeable techniques like data diddling, or to steal credit card numbers and other personal information to use themselves or more probably to sell on the dark web.

Thus hacking moved more fully into the criminal realm.

And that's where we are today. Criminal hackers have been organized into criminal organizations. They are set up with hierarchies and teams, just like their law abiding counterparts. The main difference, of course being the nature of their goals, objectives and rewards.

Today's criminal hackers have all the skills, all the experience all the tools of the best outfitted Security Operation Centers.

Therefore defending against their attacks requires continuous improvement at the highest scale. And that means improving leadership, improving technical staff, improving tools.

Hackers have the luxury of relying on our dependence upon technology. As technology, and the knowledge it provides, increases in importance to our organizational operations, the value and number of available targets increases.

In 2021 the United States (and the entire world) saw the highest ransoms asked and received for ransomware relief in the history of forever. These attacks were perpetrated by the same Russian hacker group using a well-known Ransomware-as-a-Service known as rEvil (pronounced "R" Evil).

In March of 2021, rEvil hit Acer, a computer manufacturing giant, and asked for a ransom of $50,000,000.

In April of 2021 Quanta, an electronics manufacturer that is part of Apple's supply chain, was hit with a $50,000,000 ransom by the rEvil service.

In June, rEvil was also used to hack JBS the largest beef supplier in the United States and for this one they received $11,000,000.

On July 2nd of 2021 rEvil was used to attack and successfully compromise 60 managed service providers and over 1,500 of their customers all over the world. The ransom initially requested was $70,000,000, but was later reduced to $50,000,000 for a decryption key that would work on all of the computers compromised by the attack.

This attack drew international anger and the international community collectively demanded that Russia take action. Mysteriously and without comment, rEvil's servers then went offline. For a while.

NOTE: While preparing this book for final publication, the rEvil servers came back online. They were again seen online on September 9, 2021.

The above hacks were made possible by a single service provided by a single criminal organization.

Additionally there have been attacks by others. In May 2021, Colonial Pipeline paid a $5,000,000 ransom to DarkSide, another Russian hacking group.

While these are very recent high-profile attacks, there are many others that happen every week that we don't hear about. Although the frequency of ransomware attacks is staying flat or even declining, the amount of the individual ransoms are growing exponentially. When you can get $50,000,000 for your effort, you don't need to mess around with as many $300 or $600 targets.

How much sophistication is required for these large scale attacks? The troubling answer is, "less than you would expect."

Many of the successful attacks rely on decades old techniques that could easily have been prevented, with the proper security techniques in place.

In chapter 3 we'll look at some of the specific tools that are used and you'll get the opportunity to actually use some of them. While the tools we discuss may seem overly technical for an executive or a manager, I have guided many such students in implementing and using these tools in my live training courses.

You will see that, even though defending networks can become an expensive proposition, many of the most effective attack tools can be obtained for little or no cost.

It is therefore critical to understand your own risk profile and understand the degree to which your company is exposed. By sharing these tactics, techniques and procedures with you, I hope to give you some tools you can use to look at your IT risk with a more discerning eye.

3

KILL CHAIN - YIKES!!!

TL;DR - ALL EFFECTIVE OPERATIONS START WITH A PLAN

Much of the cybersecurity vernacular is borrowed from the military or martial disciplines. This sometimes results in phrases that may at first seem overly belligerent or combative.

The term Cyber Kill Chain, first coined by the cyber security team at Lockheed Martin in 2011, is one such phrase.

The following words, written in the early 1990's by General Gordon R. Sullivan, suggest why that is.

> Never before have armies been challenged to assimilate the combined weight of so much change so rapidly. In this environment, the payoff will go to organizations which are versatile, flexible and strategically agile, and to leaders who are bold, creative, innovative and inventive. Conversely, there is enormous risk in hesitation, undue precision and a quest for certainty.

You can replace the word armies in the first line with cyber defense teams, and the quote fits us perfectly.

Admiral William Owens, opened his 1995 article, The emerging system of systems (1995, Owens) with that quote.

Admiral Owens went on to describe a method of systematically analyzing the structure of the systems involved in the attack process.

Military strategists in the United States Air Force came up with the term Kill Chain in the last decade of the previous century and the first decade of this century to describe a sequence of events that occurs when a target is attacked. The idea was that this series of steps would represent an unbroken chain of events that must take place for an attacker to be successful. If any of these links are broken, the attack can be stopped or mitigated.

In July of 2000, John A. Tirpak(2000, Tirpak) wrote an article for Air Force Magazine entitled Find, Fix, Track, Target, Engage, Assess. He credited the genesis of this idea to Gen. Ronald Fogleman, Air Force Chief of Staff and a symposium talk he gave in 1996. Gen. Fogleman's original idea of Find, Fix, Track, Target developed over time into the six phase chain of Find, Fix, Track, Target, Engage, Assess, also known as F2T2EA which Tirpak wrote about.

Having an awareness of this attack process provides defenders with some specific objectives that they can focus on in order to disrupt an attack. It also helps attackers better understand where they need to apply operational security involving our own attacks as well as our assets.

Various models grew out of that original idea, to address different situations.

One of those models is this four phase chain consisting of the following:

1) Gather information about potential target
2) Deliver tools and weapons required to the target
3) Make final Go/No Go decision to attack
4) Execute attack on target

Eric Hutchins, Senior Cyber Intel Analyst for Lockheed Martin, and the Enterprise Business System (EBS) team at Lockheed Martin adapted the Kill Chain concept to cybersecurity in 2009.

The seven phase model developed by Hutchins, et al. consisted of the following phases:

1. Reconnaissance
2. Weaponization
3. Delivery
4. Execution/Exploitation
5. Installation
6. Command and Control (C2)
7. Action on Objectives

We'll look at each of these phases of the Cyber Kill Chain in the following pages of this chapter. We explore some of the specific tools and demonstrate the ease with which these tools are used to attack your system. In the following sections of this chapter, we'll follow the activities of a fictional hacker group or 'crew' from their formation and organization through first big money attack.

Appendix A at the end of this book contains links so that anyone reading this book can explore these tools on their own.

One final note about the content. The tools mentioned in this book are easily accessible. While they can be safely used on your own personal computers and networks, especially on the virtual networks that you will be shown in this book, the misuse or abuse of these tools on system that you do not own could be a violation of laws and could incur severe penalties, including fines and imprisonment.

The information in this book is here for educational purposes only.

4

HAXORSRUS

TL;DR – THE EVOLUTION OF AN ATTACK

NOTE: *What follows is a fictional account that is a realistic representation of actual situations. The names of people and businesses are fictional and any similarity to actual people or business is purely accidental.*

A professional criminal organization in Sweden involved in old school extortion and drug trafficking has decided it's time to create a new revenue stream. Their current sources of revenue are becoming less profitable due to the increasing sophistication of law enforcement.

They decide they need to diversify into a line that is lower risk and has higher potential for profit. They have seen news reports of multi-million dollar ransoms being paid in bitcoin, and decide that's where the future lies.

They assign William, a trusted soldier, family member, and business manager, to create a hacker group and hire some of the most talented hackers he can find. William calls the group HaX0RsRUs, pronounced "Hackers – Are Us". These hackers will have a diverse set of skills. There'll be a CISCO expert, maybe a Palo Alto expert, a Microsoft expert and finally, a Linux expert. Most of them should have been experts in their particular specialty before they turned to hacking.

He assembles a group of experts. He also contacts a number of very intelligent, very talented young kids that are anxious to use their skills to get the best of "the Man" (i.e. anyone in authority). These interns will help with some of the grunt work. He also tapped the services of an out of work actor that has a graduate degree in psychology, and a young sociology major.

With this group William has the team he needs to launch this enterprise.

William calls the team together for the first time at Restaurang Istanbul, a small café in Karlskrona with heavily shellacked black wooden tables that were made locally. There's a huge clock on the far right wall as you enter the café. Seriously huge. At least 1.25 meters. It catches the eye of every person that walks through the door, and isn't staring at their phone. William sits on the opposite side of the café so he can watch everyone as they come in.

William prefers to meet people outside in the open-air back patio, but there is a light rain falling today, so it's more comfortable inside.

It is late June and most people are out and about enjoying the Midsummers Day holiday.

As they enter the café, they find William, introduce themselves, order a round of beers and immediately begin to size each other up. They're looking for some evidence that the people that they're expected to work with have some clue as to what they're doing.

After only a few beers, and some pertinent questions and excessively detailed technical responses, they develop a camaraderie, recognizing each other's competence and expecting that they will be able to work well together.

William, briefly explains the purpose of this new organization, omitting some of the grittier details for the moment, to protect the not-so-innocent.

William provides them with the address of temporary offices that they will be using for the initial setup, organization and planning.

They leave the café and head home to rest up before starting work on the following morning.

Reconnaissance

The first step for HaX0RsRUs in attacking a target is finding a target that can lead to the achievement of their main business objectives, which is to generate sufficient income to payback their investors, pay their labor and other expenses, and have enough profit to reinvest in the business to upgrade skills, equipment, and techniques, and maybe provide a little pocket change for the occasional soirée or company pizza party.

Reconnaissance is a multi-layered process, beginning with a 30,000 ft. view of the target environment. They take two approaches. The first approach is to search for vulnerabilities and evaluate whether they have the potential to lead to a victim that can afford the target ransom. The second is to look at specific industries or companies and see what vulnerabilities may be available to exploit.

For this level of reconnaissance they can start with some tools that are readily available to anyone with access to a web browser. Two of the many useful websites that are favorites of our recon team are http://shodan.io and http://osintframework.com.

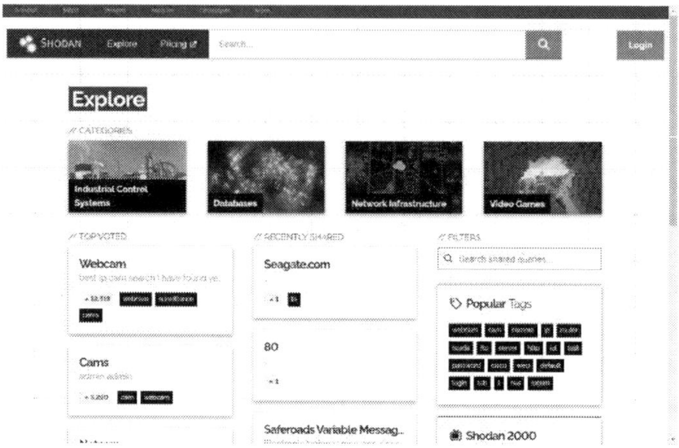

Figure 1 Shodan.io is an open source intelligence tool that exposes open ports that are publicly accessible on the internet. Some of these ports reveal potential vulnerabilities.

Shodan.io is an online database that collects banner information from ports or services that are open on the internet. Banner information is the response that is returned when an attempt is made to connect to that service. For example, when you open a browser and go to a website, you will probably receive a banner that says something like. "HTTP/1.1 200 OK" or it could say "HTTP:/1.1 401 Unauthorized".

The banner will also include other information that you won't see as part of the website, but your browser uses that information to show you the proper webpage. Sometimes information that is leaked in the banner can provide clues to possible vulnerabilities, or weaknesses that could be exploited to gain access.

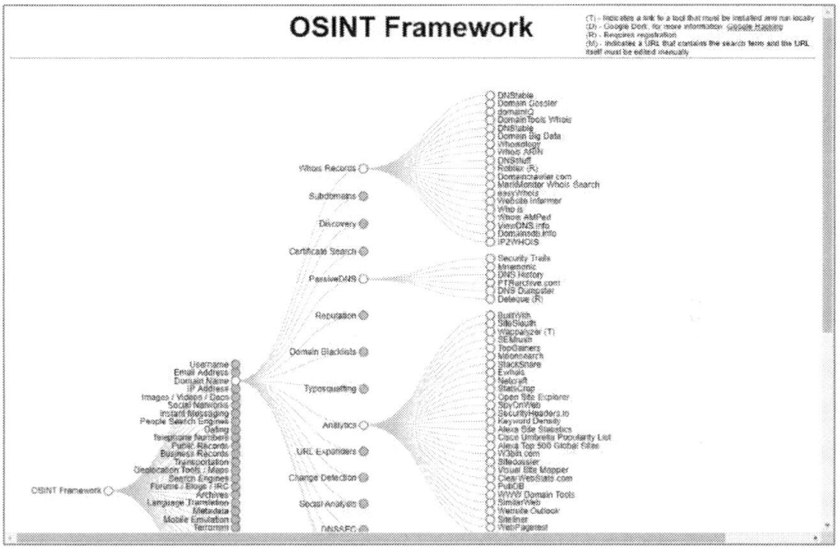

Figure 2 OSINTFramework.com is a collection of links to open source intelligence tools and services.

Http://osintframework.com is a collection of links to many websites and resources. Like any collection of links to external websites on the internet, some of the links won't work. But the ones that do work provide a tremendous amount of information. This includes information about networks, about companies, and even about individuals.

Some links lead to websites that require a subscription, but most links lead to websites with significant information at little or no cost.

Since HaX0RsRUs is a new enterprise, conserving expenditures is a top priority for the William, the business manager. He has asked the team to use only free tools for the first three months.

If they find tools that have a cost but could benefit them in their work, they are to keep a log of desired purchases. At the end of the three months, they can re-evaluate, reprioritize, and provide management with a justification of the most desired tools.

They all go to Virtualbox.org[1] and to Kali.org[2] and download the most current versions of Virtualbox and Kali. Most of them will run Kali in a virtual environment, rather than dedicating an entire physical machine to it. William says it's like everyone has their own private cloud. Sure, whatever.

Kali has the latest and greatest set of free tools that can be used throughout all phases of this operation, and they've decided that it makes sense for everyone to be using a consistent set of tools, as much as possible, so that they can more easily share information.

They begin with a basic web browser and get on the internet. Lars finds that Shodan.io has revealed that there are some PLC devices that are publicly accessible in Sudan, a small African country that currently has an

---

[1] http://virtualbox.org
[2] http://Kali.org

unstable government. PLC, or Programmable Logic Controllers, are used in Operational Technology and industrial applications. They are usually small, special purpose devices that provide some sort of critical control, like maintaining the temperature of critical manufacturing equipment, controlling the pressure in a high-pressure system, maintaining a safe speed of certain motorized equipment, etc.

Interference with a critical PLC can cripple a plant potentially costing the owners millions of dollars or an equivalent in reputational cost. For this reason alone, there is typically excellent security around them. The fact that they have found this particular controller exposed to the internet is an indication, that the security on that network could be suboptimal.

From the information available on Shodan.io they know the country and the IP Address, but they can't figure out much more than that.

So they take what they know over to the OSINTFramework.com to see what else they can discover. There they find an interesting tool call Domain Dossier and use the IP address they found at Shodan.io. They discover the equipment belongs to a network that is owned by an oil company. This seems like a positive development.

After looking closely at the PLC that they originally found, they decided that it wasn't really vulnerable to direct attack. They turn their attention to the administrative offices of the oil company, using other tools from the OSINTFramework, Google, Facebook, and LinkedIn, they begin to build dossiers on the employees, managers and executives of the company.

Odin, the out of work actor with a Master's degree in Psychology, and Luca, who spent several years working with an oil company, decide to adopt two different personas, one to develop a personal connection with a long-time HR Manager with the company and another to develop a 'professional" connection with a supervisor of operations.

Using email and fake social media accounts they begin to develop trust relationships within the company.

Freja, the sociologist student on the team, researches information about the current culture in the region and uses this information to develop profiles of the employees as well as profiles for the new personas for the team.

Lars, the Linux guy, researches the company and finds that it's a relatively small oil company compared to the other South Sudan oil companies, but its revenue is still a significant portion of the government's revenue.

They conclude that this would be a good place for their first target, because the company is not likely to have the most sophisticated defenses and the impact would be sufficient to make some headlines and earn them some street creds.

Since a national government will become a victim of a successful attack, they decide they need to do a full review of their own operational security.

They create a new shell company to rent fortified space in Karlskrona. William talks it over with his "boss" and they decide to house their offices and equipment in a local secure underground bunker that was built during the Cold War.

Noah sets up a Public Key Infrastructure so they can use strong encryption with public/private key pairs to secure their own machines, communication, and identities.

They set up multifactor authentication on all their systems, including entrances to their bunker and offices.

Lars, Noah, the Microsoft expert, and Lucas, the CISCO expert, work together to find a vulnerable public school that will soon be closed for the summer. They will compromise poorly secured systems in the school to create a relay so that their attacks appear to come from that school, rather than their actual offices. They find a school in South East Asia that they can use. They compromise several systems and install their own security measure on those systems to keep them running. They also find a backup site in a different company and do the same thing. This way, if their primary attack site goes down, they can easily switch to their secondary site to continue their attack.

They find their schools, and find potential vulnerabilities that can easily be exploited that will allow them to place their backdoors.

In the meantime, while they work on preparing the specific attacks for the schools, they set up external monitors on the sites to ensure that they are alerted if they go offline before they have control.

At the same time, Lucas identifies some alternate sites in other locations that can be used if necessary. Fortunately, public schools provide a target rich environment for poorly secured systems.

Weaponization

Their reconnaissance efforts have yielded several targets that now require a variety of weapons that need to be customized and delivered to the appropriate sites.

The targets are:

1. School Number 1 in Seoul
2. School Number 2 in Singapore
3. Social Engineering attack on Andrew (HR manager)
4. Social Engineering attack on Mohammed B. (assistant plant manager)
5. Encryption and Exfiltration attack on data
6. Operational Technology Network information acquisition.

While nobody on the team actually thinks much about the so called "Cyber Kill Chain", their actions will follow the process very closely. In fact, in this case, they will follow it multiple times. First as they establish their attack platform in multiple locations, and then as they make their final assault on the main target.

The schools' basic requirement is a remote access toolkit, but that, alone, isn't enough. They need to establish a heartbeat between the two schools. A short communication from each school to the other to confirm they are still operating as expected. If a heartbeat fails, alerts will need to go out, and where possible, the school that's still operating can pick up where the other left off.

They also need to have specialized toolkits delivered and installed that can be used against the real targets.

In order to accomplish this, they will need to gain root access to ensure they have the permissions required to install the necessary software.

As part of their own operational security, they also know that once they have exploited a vulnerability to gain access to the schools, they will need to patch that vulnerability and probably others to ensure that some other hacker group doesn't come in behind them and interrupt their plans.

During the detailed NMAP scan of the school computers, they found over a hundred potential vulnerabilities that could provide them with the access they need.

However, they know from experience that vulnerabilities that are revealed in initial scans don't always bear fruit, so they will most likely need to go through multiple attempts.

Lars sat down, prepared for a long arduous digging process that he knew would eventually lead to access.

Minutes later...

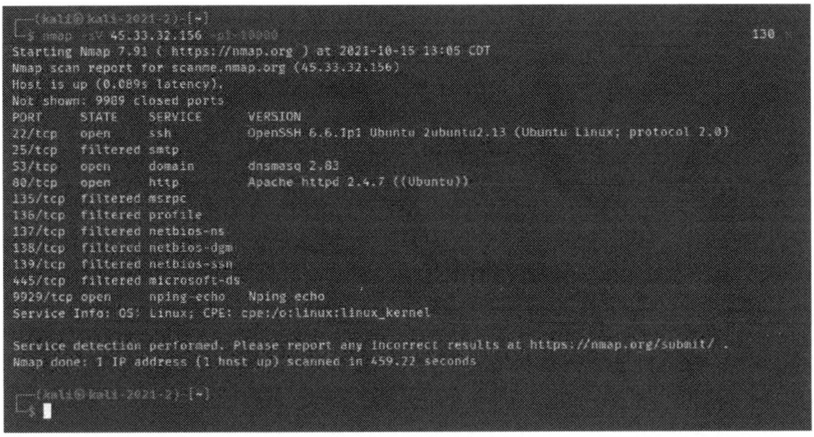

Figure 3 NMAP is a network tool that finds and identifies open ports on a network device. It is possible the most common reconnaissance tool, available for multiple operating systems. See http://nmap.org

"In-freakin'-credible!" Lars exclaimed out loud as he worked on the schools' systems. He found a website that was partially developed by a student three years ago. It was just a class project that had been created then forgotten about. Lars was guessing the kid was probably 12 years old at the time. The student obviously thought that Security through Obscurity was actually a good principle. Fortunately for Lars, that was the ONLY principle of security that was being used. The student actually created a form for uploading pictures of textbooks... and anything else that anyone wanted, it was a trivial matter to discover where the uploaded file was written.

Lars immediately went to the Kali machine and used the Metasploit tool called msfvenom to create a php file that would connect back to his machine and give him full access to the computer.

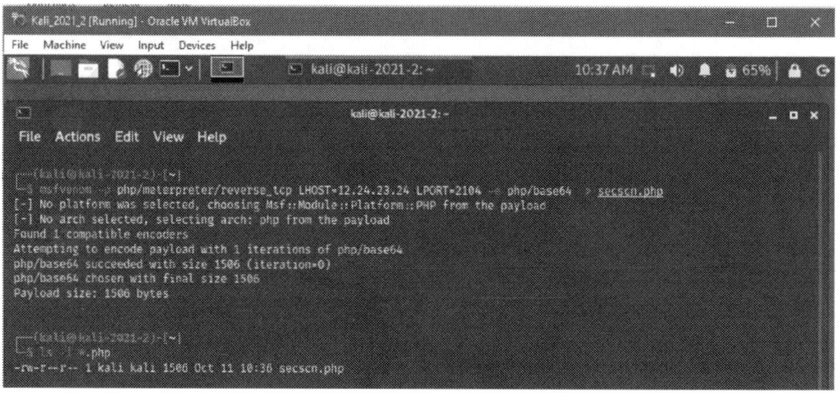

Figure 4 msfvenom is part of the collection of tools that is part of the current distribution of the Kali toolkit.

Within three minutes, Lars had created the file, uploaded it and was setting up the options in msfconsole.

As soon as the options were set, Lars opened a browser, directed it to the student's web site and retrieved his newly planted page. Boom! Connection established!

Approximately 4 minutes after discovering the insecure upload form, Lars issued the getsystem command in the meterpreter shell, acquired root permissions, and had full control of the schools' computer.

"Holy crap! I could wrap up our entire attack platform before dinner." Lars shouted to no one in particular, hoping that the Singapore school would be as simple.

As it turned out, nobody was actually listening. Which works out because Lars didn't actually care that nobody was actually listening.

Lucas and Noah were focused on learning what they could about the company and the operational technology they used. They were looking for suppliers or contractors that might have done work for the company.

They found that in that community there are only a handful of people that have technical experience working inside plants of this nature. And only a few that are previous employees of this plant. The company places a high value on its employees and therefore have an extraordinary retention rate.

They turn over all of the details they've discovered about the vendors and the current and former employees to Freja and Odin so they can decide which of these seem like the most likely candidate for providing information and then create a method or process to collect that information.

Odin and Freja thank them, then turn back to working on the human targets within the oil company. Freja has learned that the annual average income in the region is less than a quarter of the income in Karlskrona, and the monthly apartment rental is almost six times what it is in Karlskrona. She feels sad for the people in the region, but relieved that she has better conditions to live in.

William walks through the office, amazed at the amount of focus everyone in his team has. He's convinced that most of them wouldn't even notice an earthquake unless it knocked their computer out. William calls Galgamarken Indian Food and orders several pizzas.

While Lars has been working on the South Korean school, Noah has been plugging away at the Singapore school. He just realized he has an unpatched Windows 7 computer. Amazing! It's like they installed Windows 7 and disabled automatic updates so this system will have the Eternal Blue. He shakes his head, wondering why people never seem to learn. Installing updates, at least since the early nineties, has been less about upgrading to new features and more about patching critical security flaws. In all systems, but especially Windows systems.

Because Microsoft Windows is so easy to deploy and learn, it has historically maintained the lead position in market shares for desktop workstations.

When you're writing malware that you intend to spray out on to the internet, you generally want to hit as many targets as you can, so instead of developing your malware to attack macOS or Ubuntu, you write to attack the biggest target area, which is typically Microsoft. That's why most of the successful attacks you hear about are against Microsoft products, not because they're always less secure, but because they're always more available and open to attack.

Noah starts to take a sip of coffee as he scrolls through his NMAP results, then just about does a spit take as he realizes that this system has ports 135, 139 and 445 open! Unbelievable! But this is why schools can be so vulnerable. Kids that don't know any better configure insecure systems and nobody knows about it. In this case, there must have been either some malicious reason the external firewall allowed these ports to be open or it could have just been pure neglect. It's possible that someone was trying to share their schools files with their home computer and had no idea what they were doing.

Or it could be a honeypot! Honeypots often pretend to be set up by extraordinarily stupid people so that attackers will try to break in. Then they can log the ip address and the attack signature and the next time that attack is used, it may set off an alert, thwarting the attempt.

Noah decides he better tread carefully and waits for Lars to give him access to the school in South Korea so he can do the next part of the attack from there, just in case he's being monitored. By doing that, if anyone is watching, it will seem that the attack is coming from a South Korean school

Lars is installing some software on the South Korean computer, but he takes a moment, creates a user for Noah with a temporary password, gives him root access and the ability to login with SSH then sends him the login info, so Lars can continue his work while Noah works on attacking the Singapore school with EternalBlue, an exploit developed by the NSA.

Noah logs in, changes his password then fires up Metasploit.

Within 90 seconds he has a meterpreter shell and has full access to the Windows 7 computer. He immediately sets about looking around the network. Now that he is inside the network he can find other computers that were not exposed to the internet. He can also determine whether the system is virtual machine or a physical machine. If it's a virtual machine, it could be an indication that it's honeypot.

Before he runs any tools that will cause noise on the network, Noah takes a look at information that is available on the machine that he originally connected to.

Since Noah has spent his working life working with the Microsoft command prompt, he issues the meterpreter command to give him a shell that looks very familiar to him.

He runs ipconfig to find the ip addresses of the adapters on the local system.

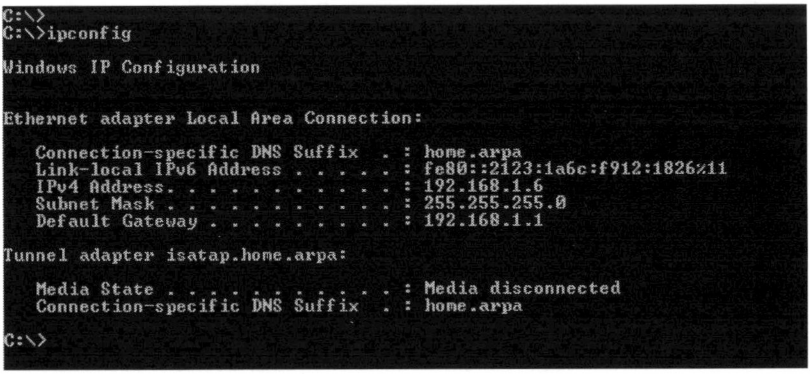

```
C:\>
C:\>ipconfig

Windows IP Configuration


Ethernet adapter Local Area Connection:

   Connection-specific DNS Suffix  . : home.arpa
   Link-local IPv6 Address . . . . . : fe80::2123:1a6c:f912:1826%11
   IPv4 Address. . . . . . . . . . . : 192.168.1.6
   Subnet Mask . . . . . . . . . . . : 255.255.255.0
   Default Gateway . . . . . . . . . : 192.168.1.1

Tunnel adapter isatap.home.arpa:

   Media State . . . . . . . . . . . : Media disconnected
   Connection-specific DNS Suffix  . : home.arpa

C:\>
```

Figure 5 ipconfig is a part of the Microsoft Windows(R) operating system used to display information about the network configuration.

Then he takes a look at the ARP table. The ARP or Address Resolution Protocol table contains a list of the ip addresses and MAC addresses that are located on the same segment. A MAC address is a globally unique identifier assigned to every port that's connected to a network.

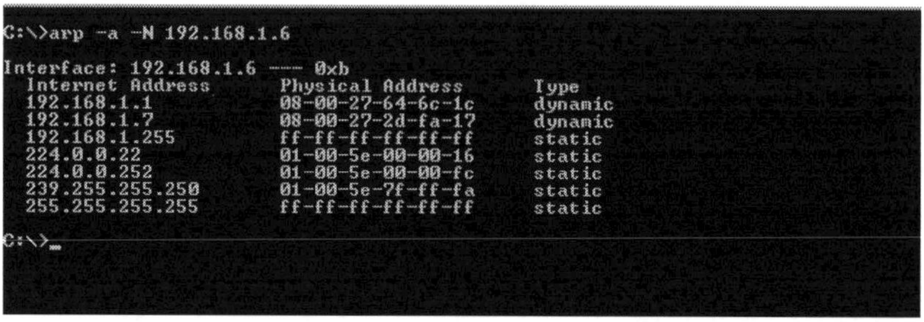

Figure 6 arp is a part of the Microsoft Windows(R) operating system used to display information about the MAC address table.

In order to have the kind of reliable access he needs, he wants to find a physical machine, possibly a server, or even a utility device that doesn't get daily human interaction. Thanks to the rapid growth of Internet of Things (IOT) there is no shortage of unsecured devices that could serve as hosts for the HaX0RsRUs attack platform.

He doesn't find anything interesting at the moment, but that could change later. He sets up a heartbeat monitor that will persist across reboots, so they can keep an eye on the reliability of this particular machine.

Once Lars gets the first attack platform setup, they'll work together to set up the same functional on the Windows systems in Singapore.

Freja and Odin have now created their fake personas and setup some social media accounts on Facebook, Twitter, and LinkedIn.

Odin researched the individuals that they are targeting at the oil company, including their social media and past relationships. The personas they've created were specifically designed to be attractive, both professionally and personally, to the targets.

They set up LinkedIn groups and Facebook groups, populated them with fake users, and then, in the course of business related discussions, they would be able to encourage the targets to participate in the discussions.

Some of the apprentices in the group can provide or generate activity in the groups.

All of this is done to make the targets feel comfortable and to trust the fake personas that they're dealing with.

While this trust is being built, Lars, Noah and Lucas work on the technical aspects of the attack.

They've already concluded that they won't be able to affect the operational technology directly, but the oil company can't be certain of that. Lucas, who has a background in oil company operational networks, will try to garner as much detailed information about their installation as possible.

When they have enough specifics about the operational technology, they'll execute a ransomware attack and in the ransom message, they threaten to destroy the plant. Lucas will make it sound plausible.

For the three months before joining this group, Lars had been engaged in writing a new variant of the wannacry virus.

On May 12, 2017 the original wannacry virus was released. Within 24 hours over 230,000 computers in 150 countries had been infected.

While the $300-$600 ransoms were modest by today's standards, the effect of the loss of machines was devastating.

Lars wrote his own shell code to use as the backdoor and added some options for propagation that will take advantage of more recently discovered vulnerabilities. It was effective in the lab and he's pretty sure it'll work at the oil company. Once he's inside the oil company's network he'll be able to adjust the code to take advantage of their particular environment.

Noah is working on a simple backdoor shell that can be embedded in a Microsoft Excel spreadsheet. The person that downloads the spreadsheet will get a security alert, so they'll have to trust the person that's sending it, which is why Odin and Freja have spent so much time on the social engineering part of the operation.

Even though the shell itself is simple to create using msfvenom, the endpoint detection system will get triggered unless he does something else. Fortunately, there are tools called packers that will obfuscate the code in such a way that it won't set off any antivirus alerts.

Noah runs the code through UPX, a simple but popular packer for malware, with plans to attach the obfuscated executable to an email.

Despite the fact that one of the oil company's PLCs had an unintended information disclosure, the rest of the company's network don't appear to have any easy access vulnerabilities like the schools did.

Therefore, the most effective way of getting access will be convincing an employee inside the company to install and run the executable.

Delivery

Once the team has gained the trust of one of the human targets, a convincing email can be crafted that will motivate the recipient to click and open the executable.

Execution/Exploitation

Odin and Freja have used their newly created personas, complete with social media groups, to gain the confidence of Andrew, the HR/Administrative Manager in the company.

They have convinced Andrew that he is talking to an HR coach who is interested at helping him out. Not only that, they'll waive their typical coaching fee so he doesn't have to risk any money!

The 'coach' offers to share with Andrew an application program that they use internally to simplify their work and make them much more productive. There's one caveat, Andrew must agree to keep it confidential and not share the program because they sell the program to their paying clients.

Andrew agrees, so they attach the program that Noah built to an email and send it to him.

Installation

When Andrew receives the email with the spreadsheet, he runs the attachment, but it doesn't seem to work. Nothing happens. He contacts the ostensible coach, who assures him, they'll have their support team look into it and get back to them.

"Our support team is swamped today, but I'll tell them you're a high priority, so they should get back to you in the next day or two", Odin assures him.

Odin looks over his partition at Noah who is now connected to Andrew's computer. "You got it?" he asks. "Sure do!" Noah said, giving him the thumbs up "Fortunately Andrew is a local administrator, so it was easy."

Meanwhile Andrew has gone back to work, completely unaware that Noah is installing programs on his system.

Noah places the programs deep in the directory structure. He needs to make sure that the programs will run automatically when the system gets restarted, so he edits the Windows registry, and creates a service using the built-in SC command. If he wanted to, he could use one of the persistence modules from Metasploit to do the same thing, but he likes to change things up and do some of this stuff "old school" occasionally, just to stay sharp.

Command and Control (C2)

Now that Noah is into the oil company and the payload is installed he wants to automate the Command and Control, so after confirming he has the necessary internet access to access IRC, he setup an IRC server and an IRC channel from which the program he installed can read its commands and post its status updates.

The channel he sets up is #SecurityTest8788. He configures his program to retrieve the commands from that channel. Now all he has to do is wait for the appropriate time, then issues the command to the IRC channel. The program will periodically check for commands. If something is there, it will pick up the command and execute. If there's no command, program will simply report the current status back to the channel.

Noah can now close his connection to the oil company from the school so any sign of that connection fades into oblivion. He won't need to reestablish a connection unless the oil company fails to report its status. Then in a few weeks, when Andrew has forgotten all about the coaching incident, Noah will send a command to the IRC channel and in a matter of minutes, the critical data of the oil company will be saved on HaX0RsRUs storage computers, encrypted, and a ransom note will be sent.

Lucas has put together what he believes to be a reasonably accurate picture of their operational technology so that the details can be shared in the ransom note.

Lucas is also in charge of setting up the DNS and the website that will hold the encrypted files. He also sets it up as an auction site so if the payment has not been received within 48 hours, the oil company's contacts, suppliers, and competitors will be notified that the data is up for auction.

Odin puts a draft of the ransom note together and shares it with the group.

```
Subject: All Data Encrypted: Recovery Info Here

Welcome

Today is not a good day for you, but I can help,
as long as you make no attempt to contact
authorities. We are monitoring your systems and
your lines of communication and will know as soon
as you try to reach out to them.

You have already noticed that you cannot access
your data. I realize that this is extremely
inconvenient for you. Your data has been encrypted
and is secure. It can only be decrypted with the
private key, which I currently hold safely.

At this point, the encryption has been applied
only to administrative and accounting services
and files. We have not yet released the detonation
of destructive software against your Valve
Automation Control Systems, your Metering Valves
manufactured by Meter Supply, and your Allied
Safety Control valves.

The software is already installed and will
detonate automatically if it does not receive the
proper hold signal from my team.

But don't worry, we will not allow this detonation
to occur as long as we continue to have your
cooperation.

Please visit the following website from your
mobile device and complete the form in order to
retrieve your decryption key using the following
access code PDYU98:
http://aaksdfopo892734iuhfasa8y98ayduf9y38u.info

HaX0RsRUs
Data Recovery Team
```

After some minor tweaking the letter is approved.

They finish setting up their infrastructure, create the website, and the payment form, Odin ends the relationship with Andrew. Once they confirmed their alerts and fail-safes are functional, they close the door, turn out the lights and take the last month of summer off.

Action on Objectives

The summer is coming to a close. All direct interaction with the oil company has been put on hold for several weeks, to allow things to settle. Odin informed Andrew that his partner did not approve of Odin providing Andrew with proprietary software without paying the $5,000 price, so he had to break off his connections with Andrew.

By now, weeks later, Andrew has all but forgotten about the brief encounter and the email attachment that didn't appear to do anything.

The team regroups in the bunker, ready to launch the attack. They issue the command, via the IRC channel to detonate the ransomware and monitor the IRC channel for status reports.

When sufficient data has been collected on the webserver, they send the email, and wait.

As a further convincer, they add another section to the web site. The new section is titled “Potential Buyers” and it lists all of their competitors, suppliers, and customers, complete with contact email addresses. Of course they obscured enough of the email addresses that would make them unreadable to anyone else who saw, but there would still be enough for the victim to recognize that they were valid.

Their company would be devastated if those people were informed of the breach, even if they didn't actually participate in the auction.

The Ransom email received a reply within 6 hours. The HaX0RsRUs team did not respond.

Then, forty–seven hours after the email was sent, the oil company visited the recovery site, paid the ransom and obtained the recovery key along with full instructions on how to recover the data. The auction was closed and the first venture by this new group was almost over.

Lessons Learned

William told them to remove all of their personal stuff from the office since it will be unoccupied for a while. He also reminded them of the importance of continued secrecy. Then they arranged to meet up the following Monday at Restaurang Istanbul.

At 10:00 straight-up, according to clock in the café, the original crew walked through the door together, Noah, Lars, Freja, Odin and Lucas.

William was sitting at a large table on the back patio. He explained they had one last thing to do before their mission was complete. He went to each person and asked them how they felt about the operation and what, if anything, they would change next time. As each person gave their assessment, William took notes.

After he had collected the relevant comments, he thanked each of them, then he heaved his backpack up onto the table. He pulled out a shoe box with 5 over stuffed envelopes. He pulled one out, read the name on the front and said, "Noah, this one is yours, there's one in here for each of you" and pushed the shoebox into the middle of the table. One by one, they took out their envelope and passed the shoebox around the table.

He also pulled out a plastic bag with 5 phones in it, saying, “If you want to contact me before next spring, make sure you use these phones. DO NOT use your regular phones!”

His tone made it quite clear that this was not a suggestion, it was a command and nobody was interested in finding out what the consequences of disobeying the command might be.

With that, they left the country for the next few months, scattering in different directions. With plans to meet back up to begin planning the next attack.

5

MOVING FORWARD

TL;DR WHAT GETS REWARDED GETS REPEATED

Ransomware attacks are too often rewarded. The events depicted in chapter 3 are not far-fetched. In fact, they happen with increasing frequency every day. There is an old maxim that says that "what gets rewarded gets repeated."[3] The tactics, techniques, and procedures used in those events are not only plausible, not only simple, but are in fact relatively easy to perform. While this may be surprising to many readers of this book, rest assured that those who are attacking our systems right now would find this information to be quite elementary. That is to say, although I have presented some technical detail, what I have shared in this book will not be much more than entertainment to our adversaries.

Today's reality is that criminal organizations are finding that ransomware attacks bring great reward. As recently as 2018, the average value of a ransomware attack was measured in the hundreds of dollars. Today the average number is in the closer to a hundred thousand dollars with the top attacks pushing into the many millions of dollars.

Most law enforcement agencies advise against paying the ransom. Often, simply remaining rational and addressing the situation calmly with the assistance of the advanced technical resources available to law enforcement is sufficient. Still, many companies panic and will act irrationally. In that moment all they can see is the pain and all they can think about is getting rid of the pain as fast as possible. And the apparent certainty of paying the ransom seems much more

---

[3] The full quote by John Jones, III is, "What gets measured gets done, what gets fed back, gets done well, what gets rewarded gets repeated."

immediate than the apparent uncertainty of a bureaucratic investigation that may or may not be effective.

The other thing to consider is that in many cases, the ransom is not the largest cost to the company. There is the indirect cost of the harm to the company's reputation which can take many months to discover. But more immediate are the direct costs of not having access to systems.

Employees, contractors and vendors still need to be paid, even when their productivity is temporarily brought to a screeching halt. And when revenue generation and collection systems are brought down, hundreds of thousands of dollars each day can be at stake.

Therefore, in the 21st century we must think differently about information security. And by 'we', I mean everyone that is involved at any level in the operation of an organization. This goes far beyond what we have traditionally thought of as 'security awareness' training programs.

For anyone, from the Board of Directors to the Truck Driver delivering the product, we all need to look at information security differently than we have in the past.

The idea that information security is something only the information security people need to deal with, is as outmoded as the idea that when you pull into a gas station, a service attendant is going to rush out, fill your tank, check your oil and wash your windows, then send you off with a big smile and a thank you.

Did you hear the 'whoosh' as that comment flew over the heads of sixty percent of the readers of this book?

Let me try that again...

The idea that information security is something only the information security people need to deal with, is as outmoded as having to buy a CD to listen to music, or record a TV show on tape if you ever want to see it in the future.

I've focused on the anatomy of a ransomware attack for much of this book, however there are many attacks that are underway right now that we also have to protect against.

Executives, managers, and leaders that will forge the way ahead into the 21st century will understand the value of information, the evolving risks that are associated with that information, and the necessity of being proactive in supporting an aggressive security posture in the organization.

Supporting an aggressive organizational security posture will be the subject of the next book in this series.

NOTES

Appendix A

Miscellaneous Online Resource List

http://shodan.io
http://osintframework.org
http://cve.mitre.org
http://kali.org
http://www.virtualbox.org
http://www.offensive-security.com
http://www.owasp.org
http://www.tryhackme.com
http://www.hackthebox.com
http://sans.org
http://comptia.org
http://isc2.org
http://CEH.org

http://DennisHouchin.com

Index

Table of Figures

References and Bibliography

Bamber, F., et al. (2017). Investigation: Wannacry cyber attack and the NHS. National Audit Office Department of Health. London.

Cole, E. (2002) *Hackers Beware.* New Riders. Singapore.

Drucker, P. (1999) *Management challenges for the 21st century.* Harper Business. New York, New York.

Harris, S., (2003). *All-in-One CISSP Preparation Guide.* McGraw-Hill. Berkeley.

Mcclure, S., Scambray, J., & Kurtz, G. (2012). Hacking exposed: network security secrets and solutions. McGraw-Hill/Osborne; London.

Owens, W. A. (1995). *The emerging system of systems.* Military Review, 75(3), 15.

Shema, M. (2014). Anti-hacker tool kit. McGraw-Hill Education.

Skoudis, E., Wright, J., Strand, Murr, M. (2020) SEC 405 Hacker tools, techniques, exploits, and incident handling. Sans Institute. Maryland.

Tirpak, J. A. (2000). *Find, Fix, Track, Target, Engage, Assess. Air Force Magazine,* 83:24 – 29, 2000. URL http: //www.airforce-magazine.com/MagazineArchive/Pages/2000/July%202000/0700find.aspx.

Internet Organized Crime Threat Assessment. (2021 August). Retrieved from https://www.europol.europa.eu/iocta-report

Kali docs official documentation. (2021). Kali.org. Retrieved From http://Kali.org/docs.

Verizon Data Breach Investigations Report. (2021 August). Retrieved from https://www.verizon.com/business/resources/reports/dbir/

Virtualbox documentation (2021). VirtualBox.org. Retrieved From https://www.virtualbox.org/wiki/Documentation

ABOUT THE AUTHOR

Dennis Houchin spent nearly four decades as an entrepreneur, software developer, engineer and top technology executive on the cutting edge in the Information Technology field. He is currently teaching Cybersecurity Bootcamp courses at New Jersey Institute of Technology and University of Kansas, University of Central Florida and others. He has also taught Cybersecurity Bootcamp courses at Vanderbilt University as an Adjunct Professor.

He has spoken to international groups such as The Project Management Institute, RSA Security, Hospitality Financial and Technology Professionals (HFTP), Toastmasters, and the Lions Club. Before setting out on a speaking career, he developed expertise in the areas in Software, Network, and Security engineering. In 2009 he began to focus more on 'hacking' human systems, specifically on understanding and applying the personality theories established by C.G. Jung to the various technical, marketing and operations people that had difficulty understanding each other. Dennis has addressed audiences of national security and hospitality industry conventions on the subject of improving IT productivity by improving the relationships between business managers and technical service professionals.

In 2015, he released his first publicly available book, The Achievement Protocol, which synthesizes the disciplines of Strategic Planning, Project Management, and Task Management into an easy to use system that anyone can use to help them achieve their life's goals.

In 2016, he published The Passion Principles, a guide to identifying and creating a life of authentic passion. He made this eBook available for free at all major eBook retailers.

In addition to teaching Cybersecurity Tactics, Techniques and Procedures to adult students and Cybersecurity Management to Executives, Managers and Leaders, Dennis is now bringing his messages of motivation, passion, and overcoming adversity to audiences around the globe. His presentations use humor and a deep personal connection with the audience to engage his audience and help them see the world anew, in a way that enables them to live a more satisfying, more fulfilling life.

Dennis teaches the same principles that he personally used to go from being a High School drop out to becoming a highly educated expert in Information Technology, completing his Baccalaureate of Science degree and earning his Master of Science degree while raising a family, building a business and creating a life worth living.

OTHER BOOKS

BY THIS AUTHOR

Contact Dennis Houchin for information on current workshop schedule or to schedule a private workshop

http://linkedin.com/in/DennisHouchin
http://DennisHouchin.com
Inquiries@DennisHouchin.com

Made in the USA
Monee, IL
12 November 2026

df350777-b0af-4cd4-95ad-5b59977ede32R01